Farm Animals

Goats on the Farm

by Bizzy Harris

Bullfrog Books

Ideas for Parents and Teachers

Bullfrog Books let children practice reading informational text at the earliest reading levels. Repetition, familiar words, and photo labels support early readers.

Before Reading

- Discuss the cover photo. What does it tell them?
- Look at the picture glossary together. Read and discuss the words.

Read the Book

- "Walk" through the book and look at the photos. Let the child ask questions. Point out the photo labels.
- Read the book to the child, or have him or her read independently.

After Reading

- Prompt the child to think more. Ask: What did you know about goats before reading this book? What more would you like to learn about them?

Bullfrog Books are published by Jump!
5357 Penn Avenue South
Minneapolis, MN 55419
www.jumplibrary.com

Library of Congress Cataloging-in-Publication Data

Names: Harris, Bizzy, author.
Title: Goats on the farm / by Bizzy Harris.
Description: Minneapolis, MN: Jump!, Inc., [2021]
Series: Farm animals | Includes index.
Audience: Ages 5–8 | Audience: Grades K–1
Identifiers: LCCN 2020021719 (print)
LCCN 2020021720 (ebook)
ISBN 9781645277071 (hardcover)
ISBN 9781645277088 (ebook)
Subjects: LCSH: Goats—Juvenile literature.
Livestock—Juvenile literature.
Classification: LCC SF383.35 .H37 2021 (print)
LCC SF383.35 (ebook) | DDC 636.3/9—dc23
LC record available at https://lccn.loc.gov/2020021719
LC ebook record available at https://lccn.loc.gov/2020021720

Editor: Eliza Leahy
Designer: Molly Ballanger

Photo Credits: Eric Isselee/Shutterstock, cover (left), 3; a_v_d/Shutterstock, cover (right); Image Source/iStock, 1; Four Oaks/Shutterstock, 4, 23tl; ahavelaar/iStock, 5; Miloszg/Dreamstime, 6–7; Jolanta Beinarovica/Shutterstock, 8–9; DonnaA Country Photos/Shutterstock, 10–11; schubbel/Shutterstock, 12; OryPhotography/Shutterstock, 13; Debra Anderson/Shutterstock, 14–15; khuntong/Shutterstock, 16, 23tr; xpixel/Shutterstock, 17, 23bl; Bilevich Olga/Shutterstock, 18; Kdonmuang/Shutterstock, 18–19, 23br; Caroline Klapper/Dreamstime, 20–21; Rosa Jay/Shutterstock, 22; oksana2010/Shutterstock, 24.

Printed in the United States of America at Corporate Graphics in North Mankato, Minnesota.

Table of Contents

Beards and Bleats

Baa!

A goat bleats.

It is morning on the farm.
The goats look for food.

Male goats are called bucks.

They have horns.

They also have beards.

horn
beard

beard

Females are called does.

They can have beards and horns, too!

Theirs are smaller than bucks'.

This doe had babies.
We call them kids.
Cute!

kid

Kids like to play!

They climb.

They jump.
Wow!

These kids butt heads!

Goats eat a lot.
They graze on grass.

They eat hay, too.
Yum!

The farmer milks the doe.
See her udder?
The milk goes in a pail.

pail

Mom is back now.

The kids are tired.

Sweet dreams!

Parts of a Goat

Take a look at the parts of a goat!

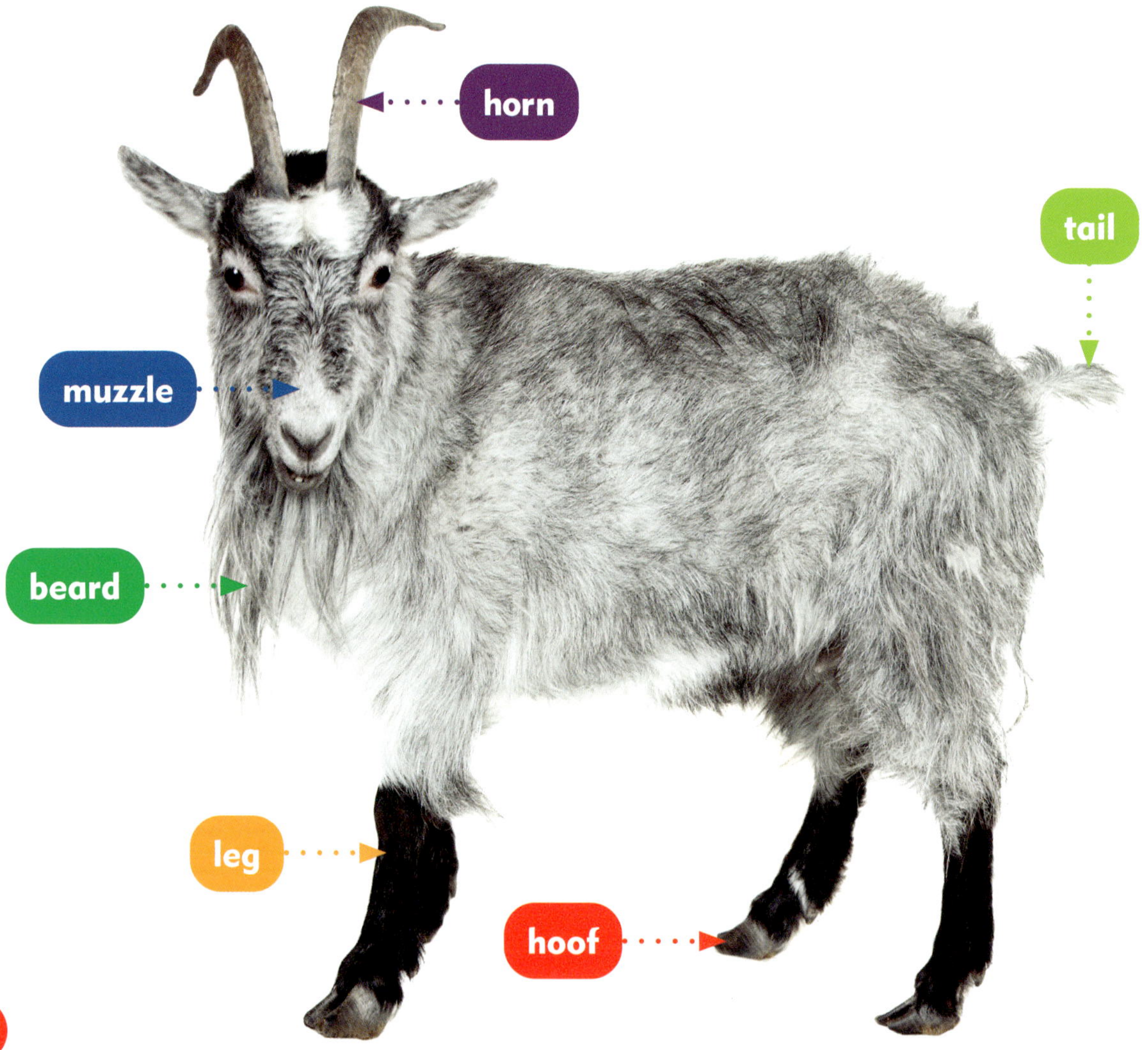

Picture Glossary

bleats
Makes the sound of a sheep or goat.

graze
To feed on grass that grows in a field.

hay
Long grass that is dried and used as food for animals.

milks
Takes milk from a cow, goat, or other animal.

Index

To Learn More

FACT SURFER

Finding more information is as easy as 1, 2, 3.

1. Go to www.factsurfer.com
2. Enter "goatsonthefarm" into the search box.
3. Choose your book to see a list of websites.